T0116345

Space Travel – the Reality

E R Niles

Photographs by D Johnson

Order this book online at www.trafford.com
or email orders@trafford.com

Most Trafford titles are also available at major online book retailers.

© Copyright 2007 E R Niles.
All rights reserved. No part of this publication may be reproduced, stored in a retrieval system, or transmitted, in any form or by any means, electronic, mechanical, photocopying, recording, or otherwise, without the written prior permission of the author.

Print information available on the last page.

ISBN: 978-1-4251-1152-6 (sc)
ISBN: 978-1-4907-8697-1 (e)

Because of the dynamic nature of the Internet, any web addresses or links contained in this book may have changed since publication and may no longer be valid. The views expressed in this work are solely those of the author and do not necessarily reflect the views of the publisher, and the publisher hereby disclaims any responsibility for them.

Any people depicted in stock imagery provided by Thinkstock are models,
and such images are being used for illustrative purposes only.
Certain stock imagery © Thinkstock.

Trafford rev. 01/09/2018

www.trafford.com

North America & international
toll-free: 1 888 232 4444 (USA & Canada)
fax: 812 355 4082

Dedicated to all who have traveled or will travel into the incredible reaches of space in search of the mysteries of our Universe

PREFACE

About the Book:

Although space travel has been a topic of great interest to many, few have an understanding of the ultimate limits that will determine its possibilities. One thing is certain. No matter the advancement in technology, whether earthling or extraterrestrial, energy requirements will determine ultimate success in space travel. True, technology will play a part. Among other things, it will determine the kinds of fuels that will be useable in deep space travel and that will, in the end, determine our successes. Are we ultimately going to be able to travel outside our own solar system to reach other worlds? Have extraterrestrial beings visited us in the past or will they in the future? These are the questions addressed in this book. It is written for the layperson, and does not require a technical background.

How to read the book:

The book begins with a short section explaining in simple terms how Einstein's Special Relativity applies in long distance space travel. Such long distance flights into space will seem possible, even one across our Galaxy in a round trip from Earth. But when we begin to consider the energy requirements of such flights, our enthusiasm and hopes for such ventures will change. In fact, the reader will come to some understanding of the upper limits that can ever be expected for any round trip travel into space. All of the fuel types including nuclear and antimatter will be considered. This limitation of success as it turns out will be very general, applying to exterrestrials as well as to Earthlings.
The technical section at the end of the Book is not intended for the lay reader, but is included for those who might want to see how the energy equations for the round trips are derived.

About the Author:

Masters degree from the University of Michigan (1957)
Professor of Physics at Antelope Valley College in California
Dean of the Math Science Division, Retired

Contents

Orion Nebula
45 minute exposure with Fuji ISO 100 Superia print film 12.5" f/5 prime focus with Lumicon coma corrector and ST-4 autoguider, off-axis.

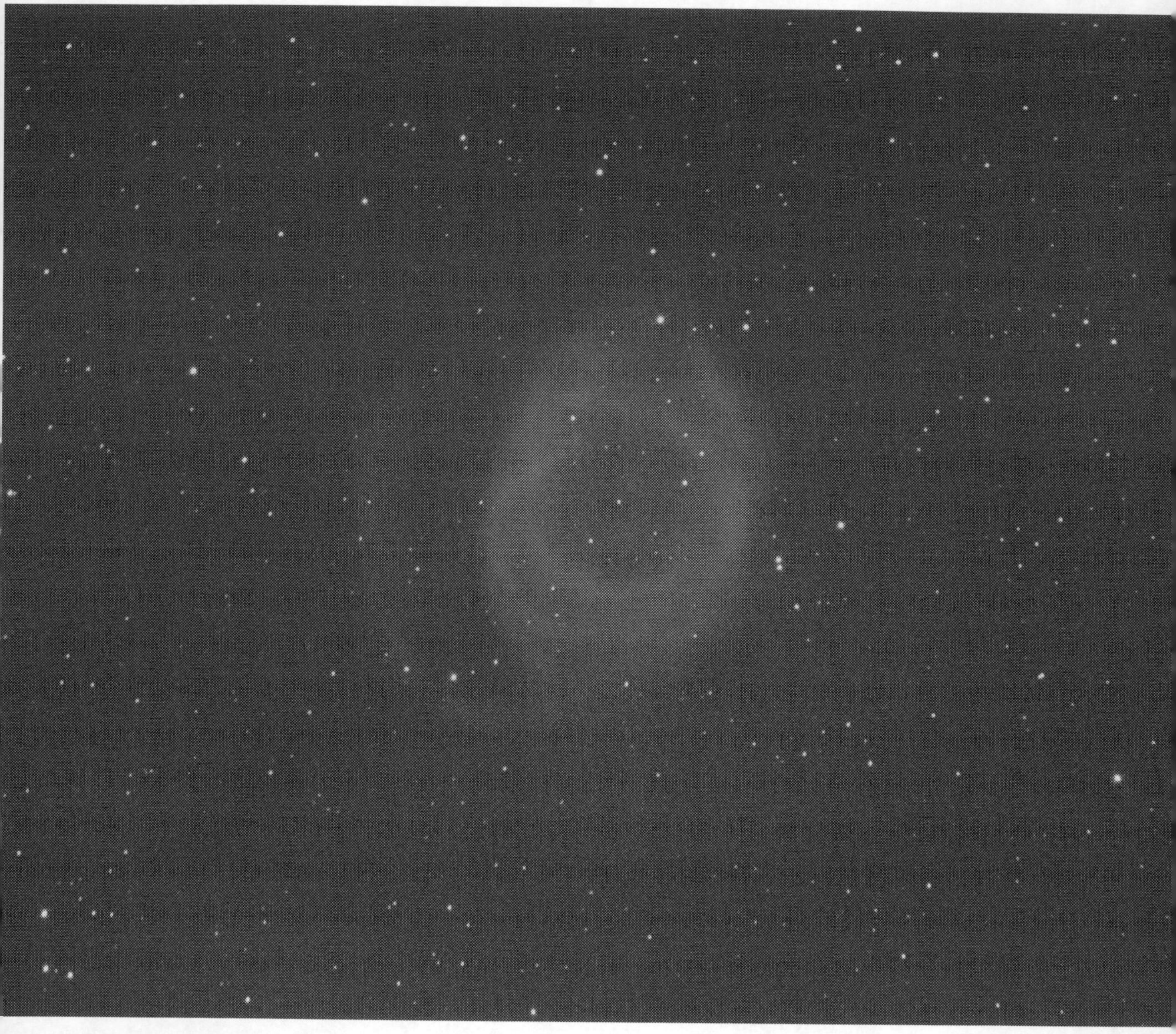

Helix Nebula

Here is a 15- minute exposure of the Lagoon Nebula on Kodak PJM Ektapress film, using a 12,5" f/5 reflector and a Lumicon coma corrector. Manually guided

Virgo Cluster

M84, M86 and other galaxies in the Virgo Cluster. Prime focus of 12.5" f/5 reflectoe using the Lumicon Coma Corrector. One hour exposure on Kodak 35mm Multispeed Ektapress print film, autoguided with ST-4 CCD. Negative scanned by Astro Photo.

SPACE TRAVEL – INTRODUCTION

Space travel has been a fascinating prospect of man, even before science has made it a possibility. Cartoons and comic strips have made it seem easy and natural. Movies make it look like the coming thing. But what would travel into space, across our Galaxy or to other galaxies entail? What would be the likelihood of such ventures? What are the limitations?

We live in an indescribably large universe. At first glance, it would seem forever impossible to traverse just our own Galaxy, a tiny part of the total Universe, because it would take about 100,000 years, even if we could travel at the speed of light, 671,000,000 MPH. Travel faster than the speed of light? It turns out that Einstein's Special Theory of Relativity predicts that it will, forever be impossible. But surprisingly this same Theory also predicts two peculiar events that aid in long distance space travel. Relativity predicts that space travelers will age less, traveling at high speeds as they move in space. That would mean that we might be able to send astronauts into space, traveling at very high speeds, who age slower than Earthlings and therefore travel farther than thought possible. But a second event, which Special Relativity predicts, is that his measurements of distances to points in space would become smaller. His space dimensions would, in effect, contract. Such an observer then, traveling at these high speeds, travels shorter distances relative to Earth's observers and has clocks that record less time to travel it, and by the same factor. This factor by which clocks slow and space contracts depends upon ones speed with respect to us as a stationary observer. Einstein showed that clocks running slow at high velocities is real, that all fast moving clocks slow, even the subtle timing mechanism of the human body that determines our aging processes. The astronaut will live longer in space than earthlings, and he will experience space contracting with respect to earthlings, by this same factor.

It should be clear however, that there is one thing that the astronaut and Earth observer will agree upon. That is their relative speed, the speed with which the two are moving away from each other. The directions are different but they both see the same relative speed. This is easy to illustrate in an example. Speed is defined simply as distance traveled divided by the time that it takes to travel that distance. If I, as an Earth observer, see you travel away from me at 100 feet in 5 seconds, your speed as I see it is 100/5 or 20 feet per second. For you as the moving observer, that distance would contract by, say a factor of 2 to 50 feet and your clocks would then run slow by this same factor. Your speed, as you see it would be 50/2.5 or 20 feet per second, in agreement with me.

This principle makes it easy to find the "relativistic advantage" that moving astronauts have over the Earth bound. It relates to the different distances traveled in space as seen by the two observers. Distance is just the velocity multiplied by the time traveled. The Earth observer, who understands classical physics, would say that the distance was just the product of the agreed upon relative speed times the time his clocks read for traveling that distance. An astronauts on the other hand, would actually travel at the same relative speed times a longer time interval because this clocks run slow. For example, if the moving observer's clocks run slow by a factor of 2, then he will travel twice the distance as Earthlings, because he will travel at the same speed but for twice as long. That factor of two then, in this example, is what we define as the "relativistic advantage" of the astronauts.

To travel into space will most likely require a spaceship designed with jet engines to accelerate, that is increase the ship's velocity away from Earth. We will need to carry enough fuel to lift off from Earth, travel to our destination, and succeed in returning home to Earth again in the lifetime of the astronauts. The goal would be to explore as much of our Universe as possible. Certainly we could hope for traversing our own home Galaxy and, maybe then beyond. We will not have the benefit of some popular mythical devices such as teleportation or engines that can explode our spaceship into the horizon in the wink of an eye. That kind of acceleration would be fatal to all aboard anyway. The trip will require four separate engine burns. First is the liftoff with the spaceship carrying along the necessary supplies for survival in space and the fuel for thee more burns. The first one will accelerate it into space to the designed velocity, hopefully high enough to take advantage of relativistic effects and take us to our distant destination. The second burn will be a reverse thrust to retard its motion in a braking fashion and essentially bring it to rest relative to, by now, far away Earth. The third burn will accelerate the ship away again, from space toward Earth; the final burn will be to bring it to rest near Earth for reentry.

Our Galaxy is an incredible place. Yet even so, it is but a speck in the Universe of millions of Galaxies. It is not a sphere but rather more like a thick disk, slowly spinning. It is made up of suns like our own, shining as brightly but so far away that they appear as point like stars at night. There are billions of stars in our Milky Way Galaxy, with the nearest one being a few light years distant. A light-year is the distance that light travels at 186,000 miles per second in one year. Not all of these suns are solar systems, that is suns with planets in orbit about them that might be habitable. The hope would be that we could find and explore solar systems anywhere in our own Galaxy and possibly, one-day even beyond our own Galaxy.

If we have been visited in the past by extraterrestrials, perhaps we can make contact or at least learn something about them. That would be our hope. There are many that passionately believe that we have been visited and describe what they believe are real-life experiences. Others doubt it with comments about the persistent lack of proof and governments deny any knowledge of such activities. As we will see, physics has something to contribute to this dilemma.

KINEMATICS AND SPACE TRAVEL

Lets look at what the Special Theory of Relativity has to say about long distant travel into space. For the moment we will disregard any energy requirement and just focus on travel itself. How far out across our Galaxy are we going to be able to travel in some reasonable time frame? That would depend upon how long we can expect our astronauts to live in space existing on only the food, water, breathing oxygen and all other supplies with which they begin their journey. The longer they can stay out in space, the greater the distance that they will travel. We should make every assumption in favor of space travel to be assured that we carry no bias against it. We can say that we are assuming the upper limits of possibility for such travel.

Let's settle on a 40-year round trip across our Galaxy, twenty years out and twenty back. We will estimate the weight of the spacecraft on the low side, a few tens of tons. That would include provisions for 3 astronauts, but exclude fuel weight. We will lift off from Earth, escape Earth's gravity and accelerate away at one "g" which humans could find tolerable. Acceleration is the increase or decrease of velocity with time. One "g" is the acceleration toward the ground that a stone would have if it were dropped from near ground. We will reach a velocity of 67,000 MPH in space for the 20 years to our destination. The results are,

> we would have traveled eleven billion, eight hundred million miles out (11,800,000,000 miles) before turning back at the 20 year mid-point. Unfortunately, that is just 0.002 light-years distance from Earth, or about 0.000002% of the diameter of our Galaxy. Our Galaxy is about one hundred thousand (100,000) light years in diameter.

There is virtually no benefit from relativistic effects at this velocity. It is clear that we must increase the velocity considerably in order to explore much of our

Galaxy. Let's increase it to 0.5 times the speed of light, about three hundred and thirty five million (335,000,000) MPH. The results,

> now we would have traveled just over 11 light-years out into space in twenty years before turning back for the trip home. That is 0.011% of the diameter of the Galaxy, still a long way from a trip across it. The astronaut's clocks run slower by a factor of 0.866 so their relativistic advantage is then about 1.155. Their space also contracts by 0.866. The clocks on Earth show that about 46 years have passed while the astronauts clocks read 40 years.

As we increase the velocity in space the relativistic advantage is beginning to show. But it is still a long way from being high enough for our astronauts to traverse the entire Galaxy. Let's try 0.99 times the velocity of light, about six hundred and sixty four million (664,000,000) MPH. Results are,

> now we would have penetrated space by about 128 light years. The astronaut's clocks have run so slowly that in the forty years that they would say they were in space, 262 years have passed on Earth. That could be as many as eleven generations.

It could be our eighth great grandson or granddaughter who would greet these venturous astronauts upon their return to Earth after their forty years in space. Still we have traversed just a little over one tenth per cent (0.1 %) of the diameter of our Galaxy. We must travel faster; try 0.9999 times the velocity of light, six hundred seventy one (671,000,000) MPH.

> Our travelers would now return after two thousand, five hundred and sixty (2,560) years have passed on Earth. Nevertheless they would be only forty years older, yet would have traveled but 1278 light years, a mere 1.28 % of the distance across our Galaxy.

Had the early Romans succeeded in launching this voyage, we could expect their return at any time now. We could ask, could we cross our Galaxy? What would it take?

Let's try .99999998c.

> Now our travelers would have nearly crossed our Galaxy, or a distance of almost 90,000 light years at least, and returned to Earth in a forty year round trip. About one hundred eighty thousand years (180,000) would have passed on Earth for the round trip.

We could ask, could we go even farther into space – travel to other Galaxies? It would seem so, except that we have not made any attempt to calculate the energy costs of our ventures, in terms of fuel mass and it's volume, and ship size including provision for a forty-year trip into space. That will be our task now.

ENERGY AND SPACE TRAVEL

In order to travel into space from Earth and return, one would first need to supply the energy to overcome the Earth's gravitational pull and then the kinetic energy (energy associated with any motion) necessary. Imagine the spaceship in a vertical position just before liftoff, ready to blast off into space. It must have enough fuel to overcome the Earth's gravitational effects and then increase its velocity to a designed value in space, carrying along enough fuel and provisions for the rest of the journey. What does it need once it gets in space? Neglecting the astronaut's needs for the journey, enough fuel to bring the ship to a halt at the turnaround point, then enough to accelerate away toward Earth once more. It must still carry enough fuel to bring it to a halt again near Earth, at the end of the journey for reentry.

The type of fuel used, whether present-day chemical or fuel requiring technologies not presently available, needs to be considered. It turns out that these energy needs are going to determine our success in space travel. For this reason we will need to consider all sources, presently available and theoretical. That would include nuclear, both fusion and fission. Fusion may be a bit more efficient than fission but it may not be as far along for space travel use technically. Both have the enormous advantage of being capable of supplying large amounts of energy, yet with smaller mass and occupying smaller amounts of space. The special advantage of these two sources of energy involves, once again, Einstein's Special Theory of Relativity. It predicts the relationship between mass (we can think of it as weight if it is confusing) and energy. In our case it means that some of the mass of a fuel can be made to disappear and be found converted into energy. How much energy can we get out of lets say, a kilogram of mass? If you could completely convert it to energy, it would be 90,000,000,000,000,000 joules. A joule is slightly less than one quarter calorie. Even though fusion or fission converts only a small per cent of its mass into energy, they are still an enormous boost over chemical technology. And it turns out that these nuclear energies are what significant deep space travel needs to succeed in any meaningful way.

Look now at possible configurations of chemical burns for space travel, applying the mysterious energy equations of Special Relativity included at the end of these pages in the technical section. Although this section is included, there is no reason to refer to it. It is only included for the scientist referral.

Assume now that we will send three astronauts on a forty year round trip from Earth, read on the their clocks. We choose only three because we need to keep the mass of the ship as low as possible so that the distances into space that we calculate are best case scenarios and the results optimum. Indeed ten astronauts or so would be more likely for a forty-year round trip.

Let the final velocity, once in space = 0.00003 x c which is about 20,133 MPH (c = the velocity of light).

> the chemical fuel required would need to be about 4,600 tons to supply the energy to bring the loaded space ship to its final velocity in space, halt it at it's destination, and then return it to Earth again in the forty year time span. The ship with provisions plus structural mass but without fuel would be about 1,100 tons. The structural mass, about 570 tons in this case, is the containment structural mass necessary to hold everything together, the astronaut's quarters, the food, oxygen, water, instrumentation, fuel and whatever else needed for the trip. Assuming a liquid propellant, the fuel could occupy, in this case, about 98% of the ship's volume. The ship's configuration with the volume necessary for this flight could be a cylinder of diameter 66 ft and length 600 ft.
>
> The lift-off burn time would be about 15 minutes, keeping the acceleration at one "g", once in space, which produces an effect on the astronauts of about what they feel as their own weight when standing on the

Earth's surface. There would be no relativistic benefit; the maximum distance out into space before turnaround in those first twenty years would be about 3.5 billion (3,500,000,0000) miles, or 0.0006 light years, a tiny 0.0000006% of the diameter of our own Galaxy.

Instead, if we were to design our trip with a space velocity of 0.000044 c or 29,500 MPH,

the fuel requirement would soar to more than 55,000 tons, the ship lift off weight could be more than 62,000 tons. There would still be no relativistic benefits. The penetration of space in those 20 years would be 0.00088 light years or more than five billion miles (5,200,000,000). The volume of our ship now has increased to what could be a tube of 160 ft in diameter and 1,100 ft long. A technical note seems appropriate here for readers who wonder how the fuel requirements are calculated. They are solutions to energy equations developed in the technical section. But what is important here is that they depend upon, among other things, the structural mass of the ship itself. The more fuel that is required, the greater is the structural mass required to survive the stresses and hold everything together, especially at liftoff. So, the greater the fuel mass, the greater the structural mass etc. More efficient fuel masses then, weigh less for a given energy output and so require less structural mass.

Clearly in the case of chemical fuels we have reached some sort of limit. We cannot reasonably expect the velocity of spacecraft in space, in a forty year round trip with three astronauts, to be greater under chemical power than about 29,000 MPH. Probably we should define an ideal upper limit of velocity for such a round trip space flight to be in the neighborhood of 29,000 MPH. Its maximum penetration of space would be about 0.00088 light years, or 5.2 billion miles for a 40 year round trip. In fact, the probability of man ever succeeding in designing such a flight, using a spaceship the size of a large ocean-going ship would seem small. We will never visit and return from the nearest star in our Galaxy, over four light years distant, with a spaceship powered by a chemical engine. So lets look at the nuclear energy possibilities.

Here are some possible configurations of fission nuclear burns for space travel. Let the final velocity in space be 6,700,000, MPH, that is 1% of the velocity of light.

> Fuel required would be about 12.8 tons of fission fuel. Ship with provisions but without fuel would be 576 tons; the fuel would occupy about 1% of the spaceship's volume; ship's configuration, with volume necessary, could be a cylinder of diameter 22.ft with a length of 100 ft. The lift-off burn time would be about 3.5 days, the thrust, once in space to give it an acceleration of "g" would be about 590 tons. There would be almost no relativistic benefits; the maximum distance out into space would be about 1,170,000,000,000 miles, about one fifth (0.2) of a light year, a tiny 0.0002% of the diameter of our Galaxy.

If we were to increase the final velocity to 107,000,000 MPH, 0.16 times the speed of light,

> the fission fuel required would be 9,900 tons; ship with provisions, but without fuel would be 1,680 tons; fuel would occupy about 10,000 cubic meters or 89% of the ship's volume; ship's configuration with volume necessary could be a cylinder of diameter 37.ft and length 330 ft. The lift-off burn time would be about 57 days, again keeping the acceleration at one g. There would be minimal relativistic benefits; Earthlings would age 40.5 years to the astronauts 40 years for the round trip; the maximum distance out into space in those first 20 years would be about 3.2 light years, 0.003% of the diameter of our Galaxy.

If we were to try our design for a velocity of 127,000,000 MPH or 0.19 times the speed of light (we will call it 0.19c), the fission fuel required would be 84,000 tons; ship with provisions but without fuel would be almost 10,000 tons. The fuel would occupy slightly more than 98% of the ships total volume; ship's configuration with volume necessary could be a cylinder of diameter 74 ft and length 640 ft. The lift-off burn time would be about 67 days. Earthlings would age almost 40.7 years to the astronauts 40 years for the trip; the maximum distance out into space in those 20 years would be about 3.8 light years, or about 0.004 % of the diameter of our Galaxy.

To accumulate 84,000 tons of fissionable material and carry it into space could reasonably (or unreasonably!) be defined as an upper limit for the forty-year round trip under fission fuel. It would appear that the limit to velocity in space then, for these round trips would be near or lower than 0.19 times the speed of light. The maximum excursion in space then would be about 3.8 light years or less. It should be pointed out, that in all of these calculations, we have assumed 90% efficiency. Only 10 % of the energy generated is not converted to the ship's motion (kinetic energy) which seems very optimistic but acceptable since we are looking for the maximum penetration of space possible. But it also needs to be pointed out here that some very difficult engineering problems are ignored and are assumed solvable. One of those is the problem of the effect of cosmic radiation on our astronauts in space for these forty-year trips.

There is another type of nuclear reaction that might be useable. It occurs when isotopes of hydrogen fuse to form more massive atoms with release of energy. If we assume here that it is twice as energetic as fission, we would find,

a flight into space at a final velocity of 0.23c, 154,000,000 MPH, under fusion power would require a fuel load of 12,600 tons for the round trip, contained in a ship of about 2,000 tons. The maximum distance out into space would be a little less than 4.7 light years, a mere .005% of the distance across our Galaxy.

We have assumed to this point that the engines of these spacecraft have been capable of producing thrusts high enough to give accelerations of one "g". In this case under fusion power, it would require a maximum thrust from the engine of almost 14,600 tons during the first burn once in space and away from Earth. If that turns out not to be possible under fusion power, we would have to lower our expectations a bit. It would necessarily compromise the mission and lower the maximum distance into space that we can expect.

If fusion engines turn out to be capable of only 1,460 tons of thrust,

that would lower the acceleration in space to 0.1g once in space, and that would lower the maximum penetration of space for the mission to about 4.2 light years from 4.7. It is due primarily to the flight spending less time at the designed velocity with its relativistic benefits on the trip out. In this case it is about 15 years instead of almost 20 years at the higher acceleration.

But let's leave that aside for awhile and continue our search for the greatest success that we can expect for fusion space flight. Assume again that we can manage one "g" acceleration. If we increase the designed velocity to 0.26c, about 174,000,000 MPH,

> it would require more than 66,000 tons of this fuel. Liftoff weight from Earth would be about 74,000 tons. It would penetrate a maximum distance in space of almost 5.3 light years. That distance in miles is difficult to comprehend for many. But if interested, it would be 31,000,000,000,000 miles. Earthlings would age by 41.4 years to the astronaut's 40 years. It could occupy a cylinder whose diameter is 66 feet and be 600 feet long. The initial thrust necessary for this flight would need to be 74,000 tons, once in space!

If the acceleration in space needs to be reduced to 0.1g, with a more reasonable 7,400 tons of thrust, the maximum distance into space would be reduced to less than 4.7 from 5.3 light years. Booster rockets would be needed to escape Earth's gravitational attraction to put our astronauts into free space.

We will probably never be able to go beyond 5.3 light years into space under nuclear power in a forty-year round trip from Earth. We can reasonably define this distance for a round trip with three astronauts as the upper limit for space travel under nuclear power. This would not be just a limit for Earth's inhabitance but for all life in our Galaxy and probably our Universe, for such a round trip under fission or fusion power. Extraterrestrial, if they have visited us or will in the future and are similar to us in size and life span, must come from no further than a little over 4 to no more than 5.3 light years from our planet Earth, under fusion power if they expected to return to their home base in a 40 year round trip. There is but one solar system, other than our own within that distance from Earth. It is Alpha Centauri at 4.3 light years distance. And it is not clear (in 2006) whether it has any planets that would be supportive of humans or other life forms, orbiting that sun.

There is an energy source which, if possible to harness and use, would be the ultimate energy source. It is a source that could allow for 100% conversion of mass into energy. It occurs in nature and in high energy accelerators when a particle and its antiparticle interact resulting in the complete transformation of the particle masses into energy. The masses disappear in effect and the equivalent energy appears essentially in the form of high-energy photons (think of them as light particles). Whether this energy would be usable as an energy source for space travel is speculative and debatable. Creating antimatter in tons and storing it for space travel may prove to be an insurmountable problem. In addition, it seems at present to be useable only for low thrust, low acceleration and so, for space travel use, long duration engine burn times. That would, as we have seen, mean less time spent at high velocities and so would dampen the full effects of relativistic benefits. But since we are looking for the greatest distance that could ever be traveled by human or extraterrestrial, we should assume it might be possible to overcome these difficulties. With tongue-in-cheek then, we assume, as before, that 90% of the energy of the reaction somehow is converted to the kinetic energy of the spaceship and that a constant one "g" thrust can be achieved.

Let's begin with a spaceship equipped with such a power source and assume a designed velocity in space of 0.5 times the velocity of light.

> The particle-antiparticle fuel required would be 550 tons; ship with provisions, but without fuel could be about 636 tons. The assumption, here again, is that there are three astronauts with provisions. If the fuel density were the same as the density of water, which includes its containment mechanism, the fuel would occupy about 30% of the ships total volume; ship's configuration with volume necessary could be a cylinder of diameter 22 ft and length 146 ft. The lift-off burn time would be about 177 days. There would be some relativistic benefits; Earthlings would age by 46 years, the astronauts 40 years; the maximum distance out into space in those first 20 years would be 11 light years, or 0.011% of the diameter of our Galaxy.

However if the maximum acceleration is only 0.1 g (10% of g), because the maximum engine thrust is low,

> the distance out in those 20 years would be reduced to less than 8.5 from 11 light years.

If the acceleration is only 5% of g because the maximum thrust which could be generated is only 60 tons,

> the distance would fall to only 5.4 light years.

Assuming that 550 tons of particle anti-particle fuel could be accumulated and contained, half particle and half anti-particle, this trip could be assumed possible.

What would it take to design a trip at 0.80 of the velocity of light or 537,000,000 MPH with this fuel?

> The fuel required would be about 25,000 tons; Ship with provisions, but without fuel would be about 3,400 tons; the fuel would occupy about 95% of the ships total volume; Ships configuration with volume necessary could be a cylinder of diameter 55ft and length 350ft. The lift-off burn time would be near 283 days. There would be relativistic benefits; Earthlings would age almost 65 years to the astronaut's 40 years; the maximum distance out into space in those 20 years would be about 25 light years.

If we design our trip at 0.81 times the velocity of light or about 544,000,000 MPH,

> the fuel required would be soar to 60,000 tons; Ship with provisions, but without fuel could be about 7,200 tons; the fuel would occupy more than 98% of the ships total volume. Ships configuration with volume necessary could be a cylinder of diameter 60 ft and length 700ft. The maximum distance out into space in those 20 years, if we could somehow put together this amount of matter-antimatter fuel, and contain it, would be a little over 26 light years, or 0.026 % of the diameter of our Galaxy. The energy for this trip would take about 13,600 times the total energy use of the entire world in 2004 (340 Quad BTU).

If the thrust using this type of fuel could be no more than 6,720 tons,

> the acceleration would be reduced to 0.1g, the maximum distance out into space before turnaround being about 13.3 light years. In other words, at

this high velocity, the space penetration would be cut in half if we could not engineer a 67,000 ton thrust with matter antimatter fuel.

We might reasonably define the velocity of 0.81 times the velocity of light and its associated penetration of space of from 13 to 26 light years, as the ultimate 40 year round trip possible for any human or extraterrestrial. It assumes that they are of the same physical size and require the same amount of food, water, oxygen and have the same life span as humans. It should be emphasized again that this does not mean that man or extraterrestrial will ever be able to accumulate this much matter-antimatter fuel, contain it, and convert ninety percent of it to energy of motion. The likelihood of life forms traveling into space a distance of twenty six light years in a round trip of forty years by their clocks, should be considered very remote. We can never traverse our own Galaxy of one hundred thousand (100,000) light years in a round trip, in which our astronauts expect to return to Earth. It should be pointed out again that the matter-antimatter power source is highly speculative. We include it here so that we can make a reasonable claim that no civilization, no matter how advanced and populated by members similar to Earth's, could travel to planet Earth from a greater distance than 26 light years and return home in a 40-year round trip from their home base. If these engines could generate no more than 5,400 tons, then the maximum distance out, for the 40 year round trip, would be about 13 light years. However, more than likely the maximum distance into space will be under fusion power for a round trip from Earth and will be in the neighborhood of 4 to 5.3 light years. And this is an unlikely, very optimistic upper limit at best.

EXTRATERRESTRIALS

If we can assume that extraterrestrial exist in our Galaxy, it would be prudent to make a special case for them since space travel might be different for them. They may be smaller than humans, consume less food and live longer. They might have advanced structural materials that are stronger than ours and perhaps lighter. We should assume anything that might be an advantage to them for deep space travel is a reality so that we can claim that our results here will show what their optimum successes in space travel are likely to be. We will assume that their space ships will be powered by the nuclear fuels, as we have for Earthlings. They may be greatly advanced, scientifically when compared to us but they cannot have propulsion systems any more potent than those that we have considered for Earthlings which includes antimatter that converts 100% of the fuel mass into energy. Hopefully we may be able to come away with an idea of the likelihood of our being visited here by aliens.

Since we have no verified witnesses of extraterrestrials, we will guess that they are half of our weight, need half of the food and other nourishment, water and oxygen. That will give them an advantage in deep space travel since the amount of mass carried along directly effects the maximum distance that they can expect to travel into space. Let us also assume that their material science has advanced to the point where the structural weight for the flight is one half of the weight of ours, or 5% of the liftoff weight, giving them a large additional advantage. Also let's allow for an eighty year round trip rather than a forty year one to accommodate a longer life span possibility. We will stick with only three alien astronauts for the crew, which is again a very minimum crew number for such a long voyage but consistent with our attempt to limit mass in order to predict the greatest successes possible.

Let us start with fission fuel and assume a final velocity of 0.16 times the velocity of light. Accelerations during the four burns for the round trip will be one “g”. With three alien astronauts and an eighty year round trip,

> the maximum distance covered before turning back would be about 6.46 light years. Earthlings would age by a little less than 81 years in those 80 years as seen by the alien astronauts. The fission fuel needed would be a little over 4,700 tons. The ship could be a cylinder 33 feet in diameter and 220 feet long. The fuel would occupy 79% of the ship's volume.

If we let the final velocity go to 0.26 times the velocity of light,

> the maximum distance covered before turning back would be a little less than 10.7 light years. Earthlings would age by a little over than 82 years. The fuel needed would be 75,000 tons of fission fuel. The ship could be a cylinder 70 feet in diameter and 627 feet long. The fuel would occupy 98% of the ship's volume.

If acceleration during the burns is limited to ten per cent of “g”,

> the distance out into space before turnaround would then be about 10, down from 10.7 light years. The reduced thrust then would be about 7,900 tons.

It would appear that we could not expect these extraterrestrial to visit Earth from a home base at a greater distance than about 10 to 10.7 light years if under fission power in these eighty year round trips.

If under fusion power, which is assumed to be twice as potent as fission, pound for pound, what would we find for this trip out with a velocity of 0.26c?

> The maximum distance traveled before turning around would be about 10.7 light years, as was the case for fission power but now the fuel weight required would be only about 8,100 tons. The ship configuration could

be 37 feet in diameter by 280 feet long.

Let the final velocity be 0.34 c., again with three astronauts aboard,

now the fusion fuel weight necessary would be about 45,000 tons. This space venture would result in travel of slightly over 14.3 light years in the 40 years before turnaround. 97% of the ships interior space would be taken up by the fuel load at liftoff.

With three astronauts again, if the acceleration were reduced by a factor 10 to 0.1g,

the maximum distance would be about 13.2 light years, down from 14.3 light years.

With a more likely ten astronauts instead of three and keeping the fuel load about the same at 45,000 tons,

the distance out to the turnaround would be about 12 light years, down from 14.3.

So, in this case at least, increasing the number of travelers from 3 to 10 has a greater effect on the mission than decreasing the acceleration by a factor of ten.

Next let's explore the effect of the lowest thrust possible on the mission. If the acceleration during the burns is reduced to the point where it is constant during the entire twenty year time of each of the two burns, outward bound, and the ship comes to rest at the end of the second burn, (i.e., the ship is accelerating or decelerating during the entire time of the flight outbound),

the distance covered would be less than 7 light years, down from an optimum of 14.3 light years. But the thrust necessary to give this 48,000 ton ship this acceleration would be only about 800 tons!

We could define about 14.3 light years distance out into space as an upper limit for an eighty year round trip, from their home base to Earth perhaps, for these three extraterrestrial beings using fusion energy. This assumes that they could design fusion engines capable of generating as much as 48,000 tons of thrust. With only 4,800 tons of thrust, the distance would be 13 light years maximum penetration of space. With 800 tons thrust, it would be about 7 light years.

Let's look at the antimatter possibilities. Assume a final velocity of 0.74c, acceleration is one "g", 3 travelers.

The maximum distance into space would be a little more than 43 light years. Fuel needed would be less than 3,000 tons contained within a ship of weight 705 tons.

With acceleration reduced to 0.1g,

this distance would only fall to about 34 light years from 43 light years. Yet the thrust level necessary would fall to just 375 tons from 3,750 tons.

If the final velocity is designed for 0.86c with 3 travelers again,

the maximum distance covered would be slightly over 65 light years or 0.065 % of the distance across our Galaxy. The matter-antimatter fuel needed would be over 42,000 tons with liftoff thrust at about 45,000 tons.

Using the above conditions again with the exceptions of the acceleration which we change to 0.1g, the trust level falls from 45,000 tons to 4,500 tons and the maximum distance falls to 48 light years from 65 light years.

Realistically, antimatter engines, if they are going to be useable at all, are likely to be capable of very low thrust. The likelihood of ever achieving a thrust in the many thousands of tons, with this type of fuel, should be considered poor. On the other hand, a thrust of 1,870 tons would give this same loaded 45,000 ton ship, an acceleration of about 0.042g. If these three extraterrestrials keep this acceleration of 0.042g constant during both burns of their trip out from their home base, their liftoff burn would be for a period of 20 years and then their breaking burn for another 20,

the maximum distance into space from their home base would then be about 21 light years. The thrust would be lowered from 45,000 tons to just 1,870 tons maximum. The distance traveled however, which measures the real success of the mission, has changed from 65 light years at optimum to just 21 light years.

With ten astronauts and about this same fuel load but a slightly higher thrust level to adjust for the additional weight of the added alien astronauts,

the maximum distance would be reduced to slightly less than 20 light years, down from 21 light years.

There is a fly in the ointment, however, regarding the design using relatively low engine thrusts. If aliens have hovered about and landed on Earth, as some claimed to have experienced, they must have escaped the gravitational pull of the Earth on their return trip to their home base. That would mean that their ship would need at least a thrust level equal to the weight of their ship plus the fuel for the two remaining burns to return home. It would define a minimum thrust level for the trip if they intend to land on Earth or hover about nearby. It turns out that the 1870 tons thrust above would be much too low for liftoff from Earth for their journey home. A solution would be that our extraterrestrials would need to park their ship in a high Earth orbit, and then launch a landing vehicle to visit Earth. In that case we could never experience the arrival on Earth of immense lumbering alien spacecraft overhead depicted in some of our imaginations or the movies.

It would seem unreasonable to claim as a maximum distance from which extraterrestrials could be expected to visit Earth to be 65 light years. More than likely, the distance would be much less, in the neighborhood of 48 light years with a lower thrust. Or even less than 20 light years with ten astronauts and a more likely very low thrust level.

THE TWO BURN ESCAPE

We may be forced, someday, to escape our planet in order to survive. We would then need to lift off from Earth and accelerate into space, carrying along only fuel for a landing somewhere on another planet. The reduced fuel load should enhance our mission's success. We would need two burns only to accomplish this mission. It differs from the slingshot, the employment of a large mass such as or black hole to accomplish the turn-around to return to Earth, because we intend to use the second burn to land at our destination. Now the question is; how far out into space would another planet have to be in order that we could reach it in, say forty years? Let's explore the possibilities.

If we accelerate away from Earth at one "g", in a forty year journey into space with three astronauts under fission power and reach a velocity of 0.2 times the speed of light (134,000,000 MPH) in a spaceship of 1,200 tons,

> it would require 5,500 tons of fuel at liftoff. The maximum distance reached into space would be about 8.1 light years.

But if we leave the Earth permanently, with no intention of returning, perhaps to colonize another planet, we need more than three astronauts. Lets increase that to ten.

> The ship would now need to be about 3,900 tons, before fuelling for the forty year trip, the increase in ship mass being attributed to the increased need for supplies water and oxygen of the seven extra astronauts and increased structural mass. Now the fuel needed for the round trip at 0.2c would be about 18,000 tons instead of 5,500 tons. The maximum distance out is still 8.1 light years.

If we increase the velocity to 0.24c (161,00,000MPH),

the ship weight with provisions, for the ten astronauts, but without fuel becomes 7,800 tons; the fuel mass needed is more than 53,000 tons. The maximum distance into space, in this forty year trip, is now 9.8 light years. What is interesting, in this case is that the maximum distance reached is much less dependent upon acceleration and therefore thrust needed, as we will see next. That generally is good news.

If we design this same flight for an acceleration during the burns of 10% g,

the penetration into space in the forty year trip is still almost 9.3 light years, down from 9.8 light years. At 5% of g, it is still about 8.7 light years. At 2% of g, it is has fallen to 6.9 light years. The thrust, once in space at this acceleration of 2% of g is only 1,210 tons, not much for the 60,000 ton liftoff weight. Of course it would need breakaway booster rockets to get it into space.

With an acceleration of one "g" again, with our ten travelers and a velocity of 0.26c, the trip would require a fuel mass of over 147,000 tons but would have little effect on the maximum distance reached. It would be only 10.7 light years, up from 9.8 light years. We could define the maximum distance out into space possible for this one way trip under fission power as about 10 light years for a spaceship transporting ten adults.

Let's go to fusion power now.

A velocity at 0.32c would require a fuel mass of almost 41,000 tons. The maximum distance penetrated in the forty years would be 13.4 light years. A velocity of 0.347c would require a fuel mass of more than 94,000 tons and a lift off mass of more than 106,000 tons. It would result in a maximum distance into space of just 14.7 light years.

This would suggest that the maximum distance out into space possible for this type of one way trip of forty years under fusion power is about 13 to 14 light years for a spaceship transporting 10 adult humans, under fusion power. In all likelihood, it may represent the maximum distance out into space that ten humans can ever expect to travel and explore a destination with no expectation for return home to Earth. Princeton has an ongoing study of solar systems around Earth with planets (the Willman Planetary System Study). That study currently shows just two solar systems known (in 2007) which are within 15 light years from Earth that contain planets. They show four planets total among the two star systems. Nothing is known as to whether they could support life.

(http://www.princeton.edu/~willman/planetary_systems/)

How far out into space would it allow us to go with our ten astronauts in this forty year, one way trip fueled by matter antimatter? If we could accumulate and contain almost 19,000 tons of this fuel in a spaceship of 4,000 tons, we could achieve a velocity of 0.9c. At one "g" acceleration,

this could result in a venture into space of as much as 80 light years. Earthlings would age by 90 years to the traveler's 40 years.

If we can only manage a lower thrust, resulting in an acceleration of 10% of g,

this distance out into space drops to about 56 light years. That thrust, once in space, would be only 2,300 tons, each of the two burns being for a period of about 8.7 years. Those on Earth would age about 73 years for the 40 years that the astronauts would say have passed, since liftoff.

It seems that a reasonable upper limit for the final velocity under matter-antimatter fuel for this type of one way space venture with ten astronauts aboard would be close to 0.9c, with an expectation of about 80 light years penetration of space. It would require 19,000 tons of matter antimatter fuel to accumulate and accomplish the feat. It would cross less than 0.08 % of our Galaxy. We would have to build separate spaceships and fuel them for each ten humans to escape Earth, if that were the primary motive. It all assumes that we could find a planet capable of supporting life, hopefully an Earthlike one within 80 light year from Earth.

This could be the matter-antimatter scenario. Astronomers find an earthlike planet at about 56 light years from Earth and we decide to colonize it. We have already engineered an anti-matter engine capable of 2,300 tons of thrust. That would give our fully loaded spaceship a minimum of 10% of g acceleration in space, once free of the Earth and our solar system. The ten carefully selected astronauts blast off and travel forty years to their destination. About 73 years pass on the Earth before we can expect that the astronauts have landed and have begun their colonization of the new planet. They were probably well into their sixties or seventies when they landed. So about 56 years after that event, and 129 years since they lifted off from Earth, we receive communication that they have landed and to send the next astronauts to continue the colonization. By the time that new crew arrives, however, the first astronauts would be long since dead from old age. No matter. We complete the colonization by sending a fleet of these spaceships, perhaps one month apart, toward the newfound home in the heavens. But to colonize the planet and establish a permanent population, families must have children. It will take a medical breakthrough for couples to have children there, after they arrive since they will be in their sixties or seventies by that time. They would need to have their children in flight so that their children would be young enough to have families of their own once established at their new home. But how slim the chances of such a venture!

This certainly emphasizes, again, the importance of the messages that we hear about the environment of our planet Earth. If it were to become unlivable, the high probably would be that we will never be able to escape to another, except perhaps to planets in our own solar system, where none seem good prospects for mass migration. If we allow our Planet's human population to increase exponentially, and the environment to degenerate and global warming to continue, we must ask ourselves; when is human existence going to become unbearable? What is the maximum number of humans that our planet Earth can support? What would life here be like as we approach that point? How much can we denude our forests, pollute our air and water, and still survive with such a population? Are we capable of surviving here, at this spot, "this blue dot" in our universe?

THE SINGLE BURN MISSION

There may be those who, some day would be willing to escape into space with no possibility of returning to our planet Earth again, or even landing somewhere on another planet, essentially a suicide decision. The mission advantage of lifting off from Earth with only the fuel necessary to accelerate into space once might be great enough to make such a venture attractive. It would make possible a much longer trip into deep space and require less thrust at liftoff.

Let's let the imagination fly now. Assume our mission is to cross our Galaxy and send pictures back to Earth. How would such a trip into space play out?

We will start with provisions for four astronauts. We probably should increase that to ten or more but we have already seen how larger crews can limit success. Matter-antimatter will be the fuel of our choice. It is the only type that will bring the space ship to high enough velocity, once in space to assure us of great enough relativistic benefits for a shot at crossing our Galaxy and sending back pictures, at least.

Here is how the trip might play out. Scientists for years have been working on the problem of engineering and building a matter-antimatter engine. Back in the year 2006, this fuel was only produced in large high energy accelerators by the microgram to milligram. Nevertheless a great worldwide effort was initiated to collect these tiny amounts and this effort has produced fuel by the ton. The seemingly impossible problems of containment with this elusive stuff have finally been solved. The magic of a new material, designed recently, will surround the crews quarters and protect our space travelers from deadly cosmic rays.

The ship rests at the Cape pointing skyward. Two large chemical fuel pods are attached to adjacent sides who's thrust will be added to the matter-antimatter engine initially to free the craft from Earth and our Solar systems gravitational reach during the initial part of the long flight. It will end only when the last astronaut dies aboard ship. Four astronauts say their last good-byes to their loved ones. Blastoff occurs; the 1,950 ton craft with four astronauts, their provisions for their remaining lifetime of 70 years, and 5,600 tons of matter antimatter fuel plus its two breakaway pods are on their way. Once in space, the astronauts settle in for their long journey. The ship's engines will burn for the over nine years. That will give the ship an acceleration of a comfortable 10% of g, requiring a maximum thrust of only about 750 tons. Eventually the mission will reach a velocity of .96c or about 644 million miles per hour.

In the first 2.33 years, the crew has traveled about 0.28 light years from Earth. Not much change! The velocity away from Earth is now about 0.24c, or over 161,000,000 MPH. Earthlings have aged slightly by 2.35 years.

In 4.66 years, they have moved 1.17 light years away from home and are now traveling at 0.48c. We on Earth are now some 73 days older than the travelers.

At the end of year seven out into space, they now have moved nearly 2.8 light years away from Earth and are traveling at a speed of 0.72c, over 483,000,000 MPH. The astronauts have aged only 7 years but their relatives on Earth have aged about 7.8 years. Much of the fuel is gone by now. The effects of relativity are felt in a big way. The astronaut's clocks have slowed and now are running only 70% at the rate of Earth's clocks. Space is contracted by this same factor.

Their headlong plunge into space continues. By the end of 9.32 years burnout occurs and the ship's engines shut down permanently. Five thousand and six hundred tons of antimatter fuel has been burned since liftoff and is now gone. Our astronauts settle down now to continue their journey at the final designed velocity of 0.96c, or 644,000,000 MPH, forever. They are six light years from

Earth now and about 9.32 years older than they were at liftoff. Their friends back on Earth, however, are twelve and one half years older. From this point forward, for the rest of their lives, they will float about aboard ship in a weightless environment.

In 20 years by their clocks, the travelers will have reached a distance of almost 43 light years from Earth. But some of us who saw the liftoff would not likely be around to celebrate that event, for over 50 years will have passed here on Earth. And we would have to wait another 43 years to receive their transmission regarding the events of that twentieth year celebration aboard ship. On Earth, that would be 93 years since liftoff. None of those who saw the liftoff would be around to witness those transmissions. Our travelers, on the other hand have aged only 20 years since they left home and would probably be in their late forties or fifties at that celebration aboard ship.

By the end of the thirtieth year, they are over 77 light years out. Back on Earth, 86 years have passed since liftoff. By the fortieth year, they are 111 light years out and have used up more than half of their supplies aboard ship. 122 years have passed on Earth.

The fiftieth year sees them over 145 light years distant from home and in their late seventies or eighties, having traversed less than 0.14 per cent of our galaxy's diameter. Earthlings, who have already aged some 158 years since liftoff by this time, would have to wait another 145 years to receive their transmissions of the event or over 300 years since liftoff. At this point the astronauts realize they have a dilemma. It would take them, by the ship's clocks, nearly another 29,000 years before arrival at the far side of the galaxy. If they expect to send information and perhaps pictures back to Earth of their arrival at that place, they must set up a timed device to accumulate and send them toward Earth, because they will all have, long ago, died by that time. So they do that, set up cameras and powerful transmission equipment on a timer, proceed to die and the ship arrives at the far side of the Galaxy. The pictures are taken automatically, by equipment that has survived 29,000 years in limbo, and the signals start back to Earth. Earthlings, by the time these pictures were taken, have aged since liftoff, equal to the time elapsed since the Neanderthals first appeared on our planet, about 103,000 years ago. And then, one hundred thousand years later, the pictures arrive at Earth. If anything is left and still readable of them, and if earthlings still exist on this dot in the heavens, they will have an understanding of what it was like there, at the far side of our own Galaxy, one of millions of galaxies in our universe in those ancient times.

SPACE STATION SETTLEMENTS

There may be a glimmer of hope for deep space travel if we could find the patience to build settlements in space that could be manned on a long time or permanent basis. They would stretch into space separated by distances of something of the order of one light year and stack in the direction of a planet or solar system that looks promising. Building the first one could necessitate a five year trip, with say 10 astronauts, capable of building of a permanent base, using the ship as a starting point. Shuttles could bring supplies, building materials and personnel to help with construction. They would be one way shuttles so that their loads would be maximized and fuel weight minimized. Each would carry another contingent of personnel plus supplies, and be designed so that they could be dismantled and reused for additional construction of their destination station. It would seem that even permanent settlements could be established, as long as the shuttles continued to operate. Since the distance to the first station from Earth would be such that the time of travel would be about five years flight time, sixty shuttles would be traveling at all times between Earth and the first settlement station. They would be staggered so that one would arrive each month. After a period of time, when the first station is functioning well, a second station could be started and established farther out in space as a second station stepping stone, as it were, into deep space. The first one could be established in six to ten years or so. The second one could also require a like period for construction if all the materials could be made available for it's shuttles to operate between the second station and first one. Once established out into space from the first station and at about two light years distance from Earth, operations and preparations could begin for the third station aboard that second one.

If we design the shuttle flights with a final velocity in space of 0.2 times the velocity of light, and assume that the trip for ten shuttle personnel will be five years time, it could require a ship weight without the fuel of about 328 tons.

That would include the weight of the astronauts and their living quarters plus their food, water, oxygen and structural mass. The fusion fuel necessary for the two burns, blast off initially and then the burn to come to rest at the destination station would be very optimistically calculated at about 760 tons. This would assume a fusion engine capable of a thrust high enough to give the loaded ship an acceleration of one "g" when in free space. That would be about 1,090 tons. If these kinds of thrusts are not forthcoming, but are limited to only 109 tons, then in the five years that it takes to travel from Earth, the shuttles would traverse only about 0.63 light years. The space station separation would then revert to this distance severely hampering success and increasing cost.

If we can build these settlements at a distance of one light year apart, but five years travel time, that would mean that the forth or fifth settlement could reach the nearest star from Earth, Alpha Centauri, at a distance of 4.3 light years. If it has a planet that looks like a good prospect, it could be explored from the advanced settlement station and pictures of that planet and of the celebration sent to Earth to arrive in about 4.3 years time. If little of significance was found there, the settlements could be expanded toward other prospects, but the difficulties will mount. The ages of the shuttle personnel born on Earth will increase as the number of stations increase. As that happens, the supply of personnel for new stations further out into space must come from the previous stations rather than Earth. In order for that to happen, the stations must become self sustaining. Families with children, some of whom will be future space travelers, must populate the stations. The shuttles must continue to operate but as time passes and the number of stations increase, they will come from earlier stations rather than Earth.

What would the first shuttle mission look like. The first astronauts and their families gather around in the mission control room. A feeling of anxiety permeates the air. Each astronaut and his or her family members know that there is a very good chance that they will never meet again. For their loved ones, the trip out will require some five years flight time; on earth, however it will appear to be about 36 days longer. A message of arrival and successful engine burn to halt the first shuttle in space will still require another year to return to Earth, but six years since liftoff. If it has been successful, the news will fly around the world in almost an instant and be greeted with jubilation.

In those intervening months since liftoff, one shuttle per month has followed, and headed out into space, a distance of one month flight time behind the other. Every liftoff has presented itself with a like scene of anxiety, fear, hope and passion. Ten more of Earth's citizens, builders, construction personal and certainly medical doctors and nurses ready to take flight, most likely forever gone from the planet Earth. As the months go by, we will begin to notice more women among the crews. Each of the stations eventually must be self support-ing and independent from earth. To be independent, the station will eventually need to nourish and support families with many of the amenities that humans need. With the arrival of a shuttle, the crew and station personnel set about dismantling it to add to the existing station. Each shuttle then adds again to the complexity and environment of that first station. Month after month, the sta-tion grows and becomes more accommodating to its human population. To be independent, it must eventually develop the capability of growing some of its own food. Nuclear fuel will no doubt be part of each shuttle's load to support the station's energy needs.

Some years after this first station's beginning, it should become more indepen-dent of the routine shuttle service from Earth. Its size would have increased many times over. Life there might even approach normalcy as its population begins to rival villages on Earth. Shops and restaurants bring some of the comforts of home. Schools, recreation areas and playgrounds appear in time. The station still is entirely enclosed from space, air tight with a magic artificial transparent material to block out harmful cosmic rays, or at least by this time a less magic solution for this problem has been found. The air within the can-opy is artificial and carefully monitored. And stations must rotate to simulate gravity.

It is very important that this independence from Earth be established. If that proves impossible as time passes, then the maximum distance into space possible by these stepping stone stations will be limited. This and the problem of insulating the stations, long term from the many hostile effects of deep space may in the end prove to be the unsolvable problems leading to the demise of such ideas.

SLINGSHOT TURNAROUND AND WORMHOLES

There is another alternative to the type of round trips into space that we have explored. It is a trip out from Earth expecting to use a "sling shot" turnaround at the midpoint in order to return home to Earth. The idea is to use a very large mass, maybe a black hole, to accomplish the turnaround. It would be an advantage over the round trips that we have considered because of the much smaller amount of fuel needed to carry along at liftoff. One problem, of course, is getting to the black hole to accomplish the turnaround. Most astronomers believe that there is a black hole in our Galaxy of the order of twenty eight thousand or so light years distant from Earth. Space travelers could travel that distance, and using the sling shot turnaround with its reduced energy requirement, return to Earth. They would then use the second burn to retard its motion near Earth for reentry.

But it would require a velocity of 0.9999998c, within a whisker of the speed of light, for the astronauts to travel 28,000 light years into space in a 50 year round trip, assisted at that distance by a slingshot turnaround. Even with the reduced energy requirement resulting from only two fuel burns, the fuel requirements would be enormous. The maximum distance out into space for three astronauts, using a slingshot turnaround, in a 50 year round trip, could be defined as about 64 light years, traveling at 0.934c, 627,000,000 MPH, at a fuel cost of 18,000 tons of matter anti-matter. To reach just 68 light years would require over 100,000 tons of this matter antimatter fuel! Extraterrestrials could not be using this method of travel, if they are cruising about in Earth orbit or landing on Earth and abducting humans, because a sling shot method does not allow any slowing or stopping in space. Two burns only, one to escape Earth and the other to come to rest later near Earth. That is what defines the slingshot.

Black holes are not the only oddity predicted by Einstein's General Theory of Relativity. Also predicted, as a theoretical possibility, is something called a wormhole. It is a distortion of space-time, as is a black hole, but it is different and never as yet observed. If wormholes actually exist, some claim that they may turn out to be a gateway to another universe or entry to another place in our own universe. How or whether these wormholes could be usable for travel to other points in our universe is speculative. But if they turn out to be real, they would have to be located close to Earth, some tens of light years away at best, so that they can be reached. One would think, since they would need to be within a distance accessible by direct observation, that we might have detected them in some way. The great chances are that, if they exist at all, that they would not be reachable by space travel from Earth and would not be an experience that would be survivable by living creatures anyway.

SUMMARY in reality, the Author's take

Under Chemical power:

No deep space travel is possible under chemical energy. It will be good for tooling about in our own solar system only. Energy wise, we should be able to reach at least the near planets of our own solar system.

Under fission power:

The probability of a forty-year round trip into space from Earth is remote but possible. It would be an extremely technical and costly venture for any nation to undertake and justify. Under fission power, it is doubtful if we will ever travel farther into space, in these forty year round trips from Earth, than about 3.2 light years. The thrust level might be limited to about 1,200-1,300 tons so that the acceleration will be limited to or near 10% of "g" in space. That would reduce the 3.2 light years to some lower figure. If, as likely to be the true, the size of the crew needs to be increased in number, the maximum distance would fall to something less. If the crew doubles to 6, it could drop to near 2.5 to 2.7 light years. Booster rockets would be needed to boost the ship away from our own solar system.

Under fusion power:

Under fusion power and a forty year round trip, we could reach 4.7 light years but more likely will manage only a trip of 4.0 to 4.2 light years, with an acceleration of no more than 0.1 g. If the crew doubles to 6, it could drop to less than 3.8 light years. Again booster rockets would be needed to boost the ship away from our own solar system. With a small crew, we might eventually be able to reach the nearest star, Alpha

Centauri in the round trip.

It should be emphasized here that fission and fusion engines, acceptable for these kinds of space travel, are apparently in their infancy technically. It is conceivable that they may never be usable in these kinds of space travel. Any engineering solutions are likely to add weight, perhaps a great deal of weight to the space ships that will reduce the success of the missions. The only thing that we can say for certain is that the energy is there and that our calculations do represent the upper limits of success with space travel. It will be up to the scientists and engineers to solve the technical problems to make it happen.

Under Anti-matter power:

Antimatter powered space flights will probably never be a reality for man or extraterrestrial. To accumulate just one ton of anti-matter seems an impossible task. Containment of matter and then separately its anti-matter also seems impossible at least in the quantities necessary for space travel. In addition, to achieve high thrusts necessary for meaningful long distant travel into space with this type of fuel, probably will never happen.

But the most important question may involve a problem with what is known as the "momentum principal". In the case of space flights, it means that we need an increasingly massive jet streamline, traveling at high speeds rearwards from the space ship in order for it to accelerate forward. With matter antimatter fuel, all of the mass is converted to energy with no mass left for the streamline except essentially photons. High energy photons do have mass so there is some mass in the streamline. But a question remains as to whether it will be enough to give the spaceship high enough acceleration for meaningful space travel? Or will we need to carry along a material such as sand to inject into the streamline to provide thrust, which would seriously effect the mission? We have assumed here that since the energy is there, it will be enough. But that may not be enough after all!

Extraterrestrials:

The energy problem would, in all probability, preclude any civilization, no matter how technically advanced, from visiting Earth. If they were located more than eight or so light years out in space, anti-matter would be the only fuel source that could be used to reach Earth in a round trip. They would have to be using the usual four-burn round trip in order to land on, or cruse about near Earth and still return to their home base. In the author's opinion, antimatter as addressed above, will never be useable for high thrust long distance travel by any civilization, regardless of talent or intelligence. If they were coming from space stations, stacked from some distant home base, and have managed in spite of all of the difficulties to reach within a few light years from Earth, one would think that we might at least pick up evidence of their communications between stations.

The likelihood is that we have not been visited in the past, nor will we be in the future unless they have originated from Alpha Centauri or nearby. A message to those that think or claim that they have had experiences of visitations in the past – Sorry!

We may someday, however, receive signals from civilizations that may be out there. Wouldn't that in a way be frustrating. We could talk to them, after we learned their language of course, but never visit them in person nor they us! Our conversations, however, would be a bit convoluted. We send a message and then, on average, according to the Drake Equation in the conclusion below, must wait perhaps 1,000 to more than 3,000 years for a reply.

CONCLUSIONS

Exploring deep space in any meaningful way is difficult to imagine. Finding a way to fuel such ventures seems to be the limiting factor. Nature herself does not present any examples of large masses moving at the high speeds necessary for space travel. Very high-energy particles of tiny masses by comparison are the only exceptions. Trips just into the fringes of our Galaxy nearby seem difficult to the point of being unlikely. It is as though nature conspires to keep us confined in our Galaxy, to our local home. In all likelihood, the maximum distance out into space that will ever be achieved by Earthlings or life elsewhere, in a 40 year round trip as clocked by the astronauts, will be 3.8 to 4.3 light years and will be under fusion power. It is not clear currently whether fission or fusion will ever be useable for deep space travel but their prospects, at present, look better than anti-matter.

All of this speaks to the issue of whether we have been visited, or will be visited by aliens from elsewhere in our universe. They would certainly be limited by the same fuel requirements for long distance space travel. They might be technically more advanced, but there could be no fuel supply better or more efficient then propulsion systems which we have considered. They represent the absolute upper limit of energy production necessary for space travel.

If we have been visited, or will be in the future, it would likely, even have to be, from our own Galaxy and from nearby. We might ask, if aliens exist and they are close by, why have we not detected them? One would think that we might pick up some faint radiation, regularly produced by a technically advanced civilization a few tens of light years away. Or such civilizations may be more rare then we would like to imagine. Recent application (2006) of the so called Drake equation which gives a best guess estimate of the number of intelligent life civilizations in our Galaxy, puts the average distance from Earth of one such civilization at about 1000 light years.

Related to this is the question of how life began on Earth and how likely it would have taken root elsewhere in our universe. Life began here, or arrived here from elsewhere by some mechanism, probably in the form of simple cells. They contained the first RNA, the precursor of DNA on Earth. Understanding the complexity of DNA with it's associated protein, and the mechanism by which it contains and uses the information it needs to create every different part of our bodies, is a cutting edge effort in biology today. This combination is, by all accounts, a microcomputer, programmed and containing data. Despite efforts to come to an understanding as to how DNA first came into being on Earth, it is still a mystery. A physicist would tend to look at it from a statistical point of view and find its computer like order and technical complexity incredibly unlikely to occur here on Earth or elsewhere by chance, in the few billion years time interval necessary. Any non life process naturally tends toward disorder. Anything we buy that is meaningful to us, a new electronic device, a toy, a new car always deteriorates over time, that is, tends toward disorder. Eventually they will be junked and in a million years or so might even be unrecognizable.

So life goes on. We may never come to find answers to these questions from the vantage point of our tiny floating planet among the seemingly infinite number of stars of our own Galaxy, itself an infinitesimal dot among all the rest of the Galaxies of our universe. What does it all mean, if anything? Science seeks an answer; the monk seeks an answer; we all seek answers.

AFTERTHOUGHT

The, what one might call "scientific" Intelligent Design, design absent of any agenda or motive, and science have some things in common. Neither has an explanation for where all this began for one. Most scientist, now at least believe our Universe began with a "Big Bang" some 13.7 billion years ago but as yet have no explanation of why or what came before that, or even whatever that question means. Some Intelligent Design advocates may say it was God but cannot explain when or how God came into existence or whatever that means.

DNA with its associated proteins, the motor of life, by any account is a computer with data that determines the physical characteristics of the animal or plant and everything else about it. In human life, the computers we build require an intelligent designer, which of course is human. Life began, after the introduction of RNA or DNA, at least on Earth, in the form of single cells. It is not a great stretch of imagination, for many, to wonder if RNA or DNA do not need a designer. The Biological community is searching for a scientific answer to this dilemma and may indeed find one eventually. The current prevailing thought apparently is that early chemistry will give an explanation for the first RNA; DNA will evolve from that. Evolution seems to make a good accounting for increased complexity of life on Earth but it requires the complex molecule of DNA itself to evolve through random changes. Computer simulations carried out at Michigan State University shows that increased complexity of DNA can occur over simulated time. But they cannot seem to start with a computer equivalent of no DNA and find that the first DNA evolves. The point is that their simulations do not seem to explain the first RNA or DNA. If biologist or chemist do succeed in explaining them through the principles of chemistry, could we answer the questions how did the delicate laws of chemistry get there – were they created in the Big Bang – did they exist before the Big Bang – if so, did they go back in time forever – if so, what does that mean?

In one way science excludes Scientific Intelligent Design from ever being viable with one rule, which although has no theoretical basis has stood the test of time and is important in keeping scientific theory free of error, bad science and even fraud. That rule states that any experiment must be repeatable. But if there were such a thing as a miraculous event, an event of a designer beyond our theories and understandings, how would it be repeatable? This could be construed as an arrogance of science and may be a basis for the irrational disgust that some scientists show for {scientific} Intelligent Design.

APPENDIX

TECHNICAL SECTION

The following pages are important only to the technical person familiar with Special Relativity principles. The equations here are used in the kinematics calculations and the energy equations following. Kinetic energy relativistically has the expression:

$$\left(\frac{m_{0s}}{\sqrt{1-\frac{v^2}{c^2}}} - m_{os}\right)c^2$$

Space is contracted for the space traveler by:

$$\sqrt{1-\frac{v^2}{c^2}}$$

and his clock rates are decreased by this same factor.
Some relationships between earth observers and astronauts are:

$$S_{er} = ut_{er} = ut_{as} / \sqrt{1-v^2/c^2} = S_{as} / \sqrt{1-v^2/c^2}$$

Accelerations are given by;

$$a_{er} = s_{er} / t_{er}{}^2 = s_{as} / t^2{}_{as} \sqrt{1-v^2/c^2} = a_{as} \sqrt{1-v^2/c^2}$$

Thrusts on the space ship are related by;

$$F_{er} = F_{as} / \left(1-v^2/c^2\right)$$

The distances during constant acceleration are related by;

$$S_{er} = 1/2 a_{er} t^2{}_{er} = 1/2 a_{as} t^2{}_{as} / \sqrt{1-v^2/c^2} = S_{as} / \sqrt{1-v^2/c^2}$$

as expected.

ENERGY AND SPACE TRAVEL

THE EQUATIONS

In order to travel into space from Earth in a round trip, one would first need to supply the energy to overcome the Earth's gravitational pull (i.e.: the potential energy) and then the kinetic energy of the spaceship and fuel needed to complete the journey, subsequent to blastoff. To return from Space, the spaceship must carry enough fuel to reverse direction so as to travel toward Earth again and then finally, enough to come to rest again near Earth for reentry. It was convenient in making these calculations to break the mid course change in direction into two parts, one to bring the ship to rest in space and then another to accelerate it away toward Earth again. That brings it to four different times in which it is necessary to burn fuel to supply energy for the mission; first, the initial burn, to accelerate away and give the spaceship, plus remaining fuel its velocity outward in space. Second, the burn to bring it to rest in space carrying the fuel for the third and fourth burn. Then the third to blast off toward Earth again and fourth, the reverse thrust to bring it finally to rest near Earth. Each burn would be different because of the varying amounts of fuel in the tanks necessary for subsequent burns. The first burn at lift off would consume the greatest amount of fuel because the ship would have to carry along the fuel for the other three burns to complete the trip. The second burn would be greater than the third and so forth.

The last page contains the equations, which describe the fuel requirements. There are five fuel variables, f_1, f_2, f_{21}, f_{22}, and f_{23}. They are individually a fuel mass of a single burn by the ship's engines. f_1 is the fuel mass needed for the first burn (at liftoff) carrying the fuel f_2. f_2 then is the mass of fuel in the tanks after liftoff and reaching final velocity away from Earth. It is the sum of f_{21}, f_{22} and f_{23}. f_{21} is that part of f_2 that will be required to bring the ship plus remaining fuel to rest in space at the turnaround point. f_{22} is that part of f_2 fuel mass needed to accelerate toward Earth again, from rest with its fuel load f_{23}. f_{23} is that part of f_2 used to bring the ship to rest finally near Earth. The fuel mass to supply the gravitational potential energy change at liftoff is negligible for most of these calculations because it is very small when compared to the fuel mass energies for accelerations and decelerations necessary for deep space travel. The total fuel requirement for the round trip into space and return is the sum f_1, f_{21}, f_{22}, and f_{23}. This same requirement should be true for aliens from elsewhere in our Universe who visit us on Earth if they expect to return to their home base.

Equation A describes the energy relationships for lift off of the spaceship using the appropriate relativistic formulation. Any meaningful space travel requires relativistic formulation because of the very high velocities necessary. The left member, f_1c^2 where c is the velocity of light, is the energy supplied the engines of the spacecraft to give it its kinetic energy in space (the second term in the right member), plus the kinetic energy of the remaining fuel f_2 carried aboard the craft (the first term in the right member). The left member assumes that all of the fuel mass f_1 is converted to energy. Fusion and fission technology that we have today involves conversion of only a portion of mass into energy. Assuming that all of the fuel is 100% converted into energy represents, in effect, an upper bound. That is, no smaller amount of fuel, assuming a different technology of some sort will be possible. The only time nature demonstrates this type of interaction is at the particle level when an anti-particle and particle interact, with 100% conversion of their masses into energy. It would be a wild stretch of reality perhaps, to assume that some day we would be able to bring aboard the spacecraft two tanks, one for a mass of substance and a second for its anti-substance, which then could be fed in small amounts from each tank into the engine chambers as our energy source. Nevertheless, that is essentially what is assumed in these matter anti-matter type fuel calculations. The equations are generally useable for nuclear power and by the classical case such as chemical power. The mass conversion for chemical is assumed to be one part in a billion (i.e.: if f_1 turned out to be 1 kg of matter antimatter, the equivalent chemical mass would be one billion kg). For fission 1% conversion to energy, for fusion 2% and matter anti-matter 100%. The relativistic kinetic energy equations, of course, reduce to the classical values at low velocities, a very convenient feature.

Equations A, B, C, D and E are five equations solved simultaneously for the fuel masses in terms of the ship rest mass and the relativistic change in mass. The total fuel necessary for the round trip is given by the equation near the bottom of the page with the left member "$f_1+f_{21}+f_{22}+f_{23}$". This equation gives the ideal theoretical fuel mass (100% mass converted) necessary to complete the round trip as a function of the rest mass of the space ship and its relativistic change in mass. The energy for the first burn at blast off is just this mass, f_1 multiplied by the velocity of light squared. To give a feel for the potency of the matter anti-matter fuel, if we apply these equations to a round trip flight into space in a 575 ton space ship at speeds of 6,700 MPH, out and back, it would require only 0.5 grams of fuel which includes the energy to escape Earth's gravity. (Escape energy is not part of these equations). It assumes that 90% of the energy generated is converted to kinetic energy, with a loss of only 10%.

The energy for a probe (out only with no fuel left after the initial burn), would be given by multiplying f_{23} near the bottom of the page, by the velocity of light squared.

The fuel mass to liftoff from Earth, then reverse thrust to bring the ship to rest in space at some point is given by the sum of f_{22} and f_{23}, second from the bottom of the page. This would be a two-burn mission and is not generally a round trip one. It can be derived by calculation similar to the process of finding the total fuel masses for the round trip. It would also be the appropriate mass for liftoff from Earth carrying only enough fuel for landing on Earth again using a slingshot turn around in which the ship might use a large gravitational mass such as a black hole to reverse direction without fuel loss. The fuel f_{23}, carried along, would then supply the reverse thrust to retard its motion for safe landing near Earth.

The " M Δm", relativistic change in mass is given by:

$$M_{os}\left[\left(1-\frac{v^2}{c^2}\right)^{-\frac{1}{2}}-1\right]$$ where the rest mass M_{os} of the spaceship includes the structural mass and all provisions necessary to sustain life aboard. This does not include fuel. Some estimates have been made in this regard including food supply, water, oxygen, quarters etc.

Assumption and observations:

1. In all of the calculations it has been assumed that the structural mass of the spaceship must be ten percent of the lift off mass. This would, presumably, be related to the maximum stress possible on the structure of the ship. That would involve the material itself as well as the design. An exception was made for Exterrestrials.

2. The potential energy to escape Earth's gravity is taken into account in the calculations although it is almost always insignificant.

3. Use of matter-antimatter as a fuel source for propulsion, especially in huge amounts, may not be possible. Since all the mass is converted to energy, no particles except high-energy photons and pions would be left to propel the ship according to the momentum principle. Some proposals that have been made (Scientific American June 2005, P 85) seem to allow only limited thrusts during long duration and inefficient engine burns. Long duration burn-times mean that the craft will spend less time at relativistic velocities – a great detriment to long distance round trip space travel. Storage of antimatter for use in spacecraft would be very difficult. If this fuel source, one day is shown to be unusable for space travel, we are left with only fusion and fission sources as the

best candidates for deep space round trip travel.

4. The efficiency of the energy transfers in these calculations has been assumed, very optimistically at 90%. In other words 90 % of the energy generated is converted to kinetic energy. Assuming 90% efficiency is consistent with an optimistic attitude toward space travel, if not a realistic one.

5. Some have postulated that a ramjet could be designed to collect or scoop up, as it were, matter anti-matter in space and that it might solve the stubborn fuel problem bugging space travel. Some scientists have argued that the scoop would have to be so large as to make that idea impractical. Another argument to consider against this might be that the engine thrusts and accelerations using this system would tend be too low for meaningful space travel. A burn producing a constant acceleration of 0.001 meters/sec^2 would take about 1,900 years to reach a velocity of 0.2c, if it started from rest. The use of light sails, a system where powerful laser beams on Earth are projected onto sails attached to the space ship for propulsion fall into this same category.

6. To determine the energy equivalent of fission fuel, it was assumed that 1% of the mass was converted to energy. For fusion fuel it was 2% converted and for matter-antimatter it was 100%. In chemical burns, it was assumed that one part in one billion of it's mass was converted to energy.

7. The distance traveled during the engine burns is assumed to be at constant acceleration and therefore is just the products of one half the final velocity and the time of burn. The relativistic benefits during the engine burns however can, in effect, make this final velocity higher. It is assumed to be the integral $\int dx/(1-x^2)^{1/2}$ = ASIN(x) where x= (final designed velocity)/c. The distance covered then is ½(Asin(v/c)) times the burn time.

8. Space travel from our planet Earth is currently is in its infancy. Technically where we will be in a hundred or a thousand years is guesswork.

Nevertheless, it is very possible that travel among the stars outside our own planetary system may never occur, at least in the lifetime of human travelers. Solving the technical problems to achieve thrust levels necessary for high enough acceleration in space seems unlikely. Finding anti-matter by the ton and learning ways to contain it efficiently and effectively aboard the spaceship is so daunting a task that one could reasonably conclude it will never happen. Still the assumption here is that science will find a way. These calculations then will give us an upper limit for space travel success. It would seem reasonable to assume that these upper limits would also apply to other life forms in our universe, if they do exist. Unproven theoretical schemes such as warp drives, inertia drives, drives requiring utilizing fuel planted into space along the route of travel and light sails driven by powerful beams from Earth, are not considered practical or realistic for manned, long distant space flight.

SPACE TRAVEL EQUATIONS energy and fuel mass

A $f_1c^2 = f_2\Delta mc^2 + m_{os}\Delta mc^2$ Blast off from Earth

A' $f_1c^2 = (f_{21} + f_{22} + f_{23})\Delta mc^2 + m_{os}\Delta mc^2$

B $f_2 = f_{21} + f_{22} + f_{23}$

C $-f_{21}c^2 = -f_{22}\Delta mc^2 - f_{23}\Delta mc^2 - m_{os}\Delta mc^2$ Come to rest in space

D $f_{22}c^2 = f_{23}\Delta mc^2 + m_{os}\Delta mc^2$ Blast off toward Earth again

E $-f_{23}c^2 = -m_{os}\Delta mc^2$ Come to rest near Earth

D' $f_{22} = (m_{os}\Delta m)\Delta m + m_{os}\Delta m$

C' $f_{21} = (m_{os}\Delta m^2 + m_{os}\Delta m)\Delta m + (m_{os}\Delta m)\Delta m + m_{os}\Delta m$

$= m_{os}\Delta m^3 + 2m_{os}\Delta m^2 + m_{os}\Delta m$

$f_1 = ((m_{os}\Delta m^3 + 2m_{os}\Delta m^2 + m_{os}\Delta m) + (m_{os}\Delta m^2 + m_{os}\Delta m) + m_{os}\Delta m)\Delta m + m_{os}\Delta m$

$f_1 = m_{os}\Delta m^4 + 3m_{os}\Delta m^3 + 3m_{os}\Delta m^2 + m_{os}\Delta m$ Blast off fuel mass

$f_1 + f_{21} + f_{22} + f_{23} = m_{os}\Delta m^4 + 4m_{os}\Delta m^3 + 6m_{os}\Delta m^2 + 4m_{os}\Delta m$ Total fuel mass round trip

$f_{22} + f_{23} = m_{os}\Delta m^2 + 2m_{os}\Delta m$ Two burn. (Derivation not shown)

$f_{23} = m_{os}\Delta m$ Fuel mass for probe or suicide mission

To make use of the equations requires extensive calculations. We need to calculate the fuel requirements for a particular trip, which depend on the designed velocity of the craft in space as well as the mass of the spaceship itself. That mass depends upon the number of astronauts, their needs for the entire trip, and the structural mass necessary. It has been assumed that the structural mass must be ten percent of the liftoff mass which, of course, depends upon all of the above. The construction of a spreadsheet is recommended.

Printed in the United States
By Bookmasters